W9-BNW-074

21st Century
Basic Skills
Library

LINE GRAPHS

Books I Read

Week 1

Week 2

Week 3

Published in the United States of America by Cherry Lake Publishing
Ann Arbor, Michigan
www.cherrylakepublishing.com

Consultants: Janice Bradley, PhD, Mathematically Connected
Communities, New Mexico State University; Marla Conn, Read-Ability
Editorial direction and book production: Red Line Editorial

Photo Credits: Shutterstock Images, cover, 1, 6; iStockphoto/Thinkstock, 4,
14; Hemera/Thinkstock, 8; Thinkstock, 10, 16; BrandX Images/Thinkstock,
20

Copyright ©2014 by Cherry Lake Publishing
All rights reserved. No part of this book may be reproduced or utilized in
any form or by any means without written permission from the publisher.

Library of Congress Cataloging-in-Publication Data
Edgar, Sherra G.
 Line graphs / Sherra G. Edgar.
    pages cm. -- (Let's make graphs)
 Audience: K to grade 3.
 Includes bibliographical references and index.
 ISBN 978-1-62431-392-9 (hardcover) -- ISBN 978-1-62431-468-1
(paperback) -- ISBN 978-1-62431-430-8 (pdf) -- ISBN 978-1-62431-506-0
(ebook)
 1. Graphic methods--Juvenile literature. I. Title.

 QA90.E345 2013
 518'.23--dc23

                                      2013004939

Cherry Lake Publishing would like to acknowledge the work of The
Partnership for 21st Century Skills. Please visit *www.p21.org* for more
information.

Printed in the United States of America
Corporate Graphics Inc.
July 2013
CLFA11

TABLE OF CONTENTS

What Is a Line Graph?

Sam wants to count his cat's treats each day. He can use a **graph**. Graphs show **data**.

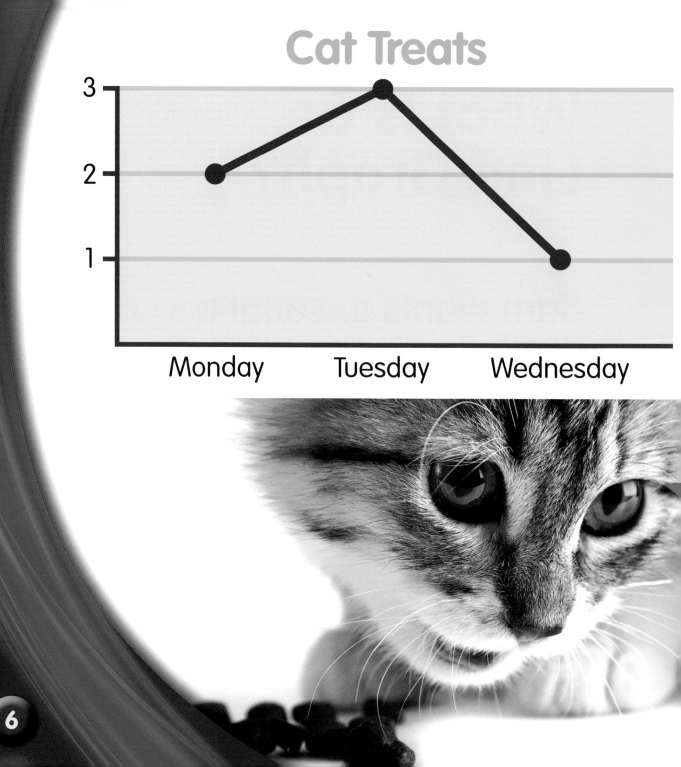

Cat Treats

3	
2	
1	

Monday   Tuesday   Wednesday

Sam made a **line graph**. Line graphs have **points**. Each point stands for an **amount** at a certain time. Connect the points to make a line.

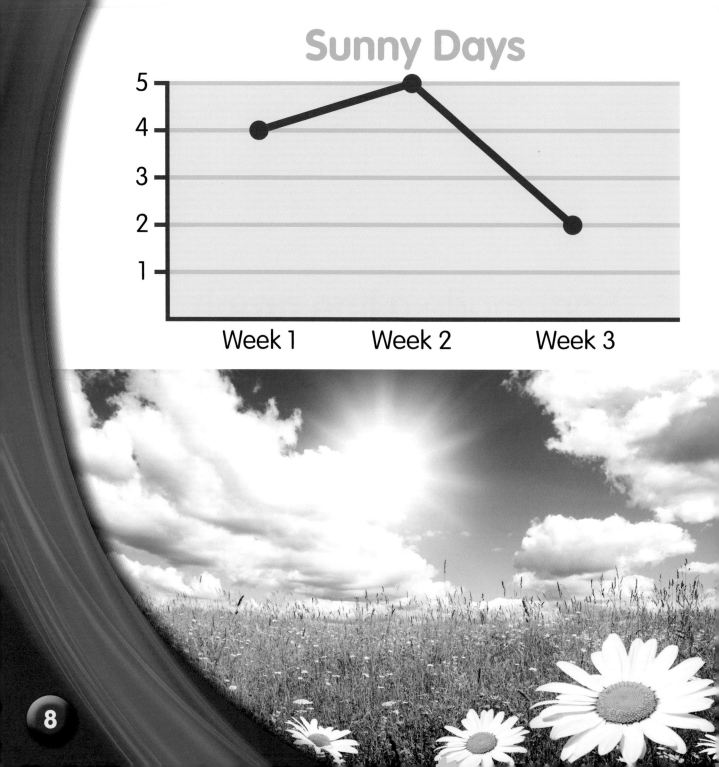

Sunny Days

A line graph showing Sunny Days across three weeks:
- Week 1: 4
- Week 2: 5
- Week 3: 2

Line graphs show how amounts change over time. The points show the data at different times.

Making a Line Graph

Sasha wants to count the fruit she eats.

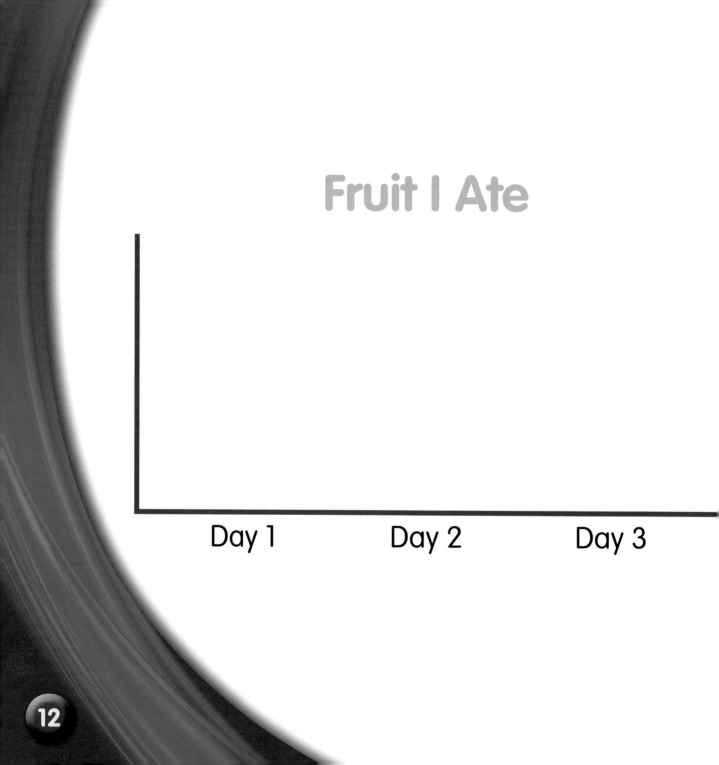

Fruit I Ate

Day 1   Day 2   Day 3

Sasha wrote days 1, 2, and 3 under an **L**. This shows the time.

Fruit I Ate

Sasha numbered 1 to 5 up the side of the **L**. This will show the fruit amounts.

Fruit I Ate

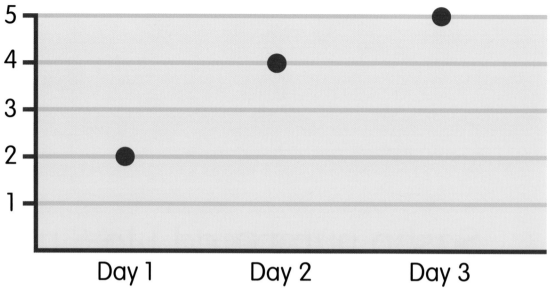

Sasha put a point at each amount. Then she made a line.

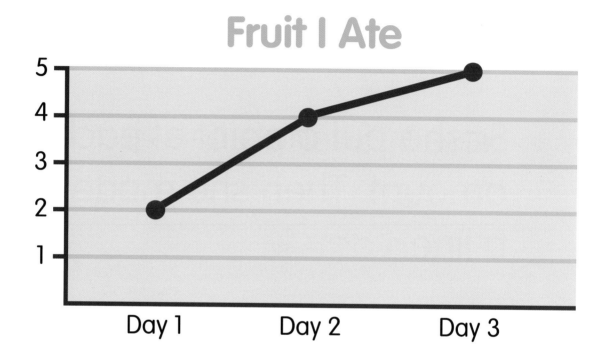

Fruit I Ate

Here is Sasha's line graph. You can see she ate the most fruit on day 3.

You Try It!

Make a line graph of dogs you see. Count the dogs you see each day for a week. How does your line graph look?

Find Out More

BOOK

Nagda, Ann Whitehead. *Tiger Math: Learning to Graph from a Baby Tiger*. New York: Square Fish, 2002.

WEB SITE

WatchKnowLearn.org
www.watchknowlearn.org
Watch videos to learn about graphs and other math topics.

Glossary

amount (uh-MOUNT) how many or how much there is of something

data (DEY-tah) amounts from a graph

graph (GRAF) a picture that compares two or more amounts

line graph (LAHYN GRAF) a graph made by connecting points to show amounts or changes over time

points (POINTS) dots that stand for amounts on a line graph

Home and School Connection

Use this list of words from the book to help your child become a better reader. Word games and writing activities can help beginning readers reinforce literacy skills.

a	each	making	this
amount	eats	Monday	time
an	for	most	to
and	fruit	numbered	treats
at	graph	on	try
ate	graphs	over	Tuesday
can	have	point	under
cat	he	points	up
certain	here	Sam	use
change	his	Sasha	wants
connect	how	see	Wednesday
count	I	she	week
data	is	show	what
day	it	shows	will
days	line	side	wrote
different	look	stands	you
does	made	sunny	your
dogs	make	the	

Index

About the Author

Sherra G. Edgar is a former primary school teacher who now writes books for children. She also writes a blog for women. She lives in Texas with her husband and son. She loves reading, writing, and spending time with friends and family.